3 Georgia Milestones Grade 3 Math Practice Tests

Full-Length Test Prep with Detailed Answer Explanations

Dr. A. Nazari

3 Practice Tests to Get You Started!

Hey there, future math whiz! 

This book has **3 full practice tests** to help you warm up for the real thing. Think of it like stretching before a big game — these tests will get your brain ready and show you what to expect!

👍 Three tests is the **perfect start**!

👍 Each one helps you feel **more ready**!

👍 You'll be surprised how much you **already know**!

Sharpen your pencil and let's get warmed up! 

> 66 Three practice tests is a great way to start. Take your time with each one, and you'll feel more confident every step of the way! 99

📘 How to Use This Book 📘

☰ What's Inside This Book

- **3 Full-Length Practice Tests** — Each one covers all the Grade 3 math topics you need to know!

- **Answer Key with Explanations** — Find out why each answer is correct, not just what the answer is.

- **Reference Pages** — A math symbols chart and multiplication table you can peek at any time.

- **A Test Tracker** — Write down your scores and watch your confidence grow!

📅 A Simple 3-Test Plan

With just 3 tests, here's a great way to use them:

- **Test 1**   **The Warm-Up.** Take this test without a timer. Get comfortable with the question types. Don't worry about your score — just do your best!

- **Test 2**   **The Practice Round.** After reviewing Test 1, try this one with a timer (ask a grown-up!). Focus on the topics that were tricky last time.

- **Test 3**   **The Real Deal.** Treat this like the actual test: quiet room, timed, no peeking at answers. See how much you've improved!

● Multiple Choice

Pick the **one best answer** from choices A, B, C, or D. Not sure? Cross out the ones you know are wrong, then pick from what's left. That's a smart move!

✏ Short Answer

Write your answer **and** show your work! Even if your final answer isn't right, showing your steps can earn you credit. Use scratch paper if you need more room.

66 After Each Test 99

Flip to the Answer Key and check your work. For every question you got wrong, **read the explanation carefully**. Then write the tricky topics on your Test Tracker page. If you need extra help, grab our **Grade 3 Math Study Guide!**

⭐ **Fun fact**: *Three tests is all it takes to see real improvement! Most kids feel way more confident after just a few rounds of practice.* ⭐

Find more at
ViewMath.com/GA-Grade3

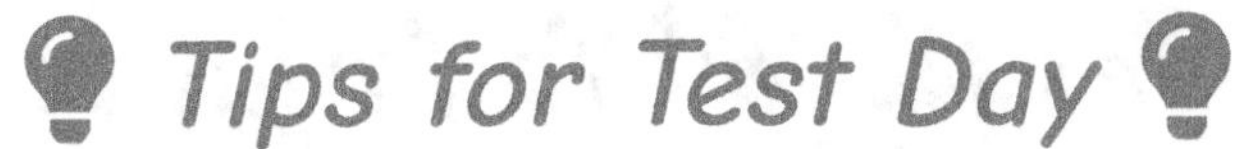

Tips for Test Day

Easy tricks to help you feel calm and do your best!

🌙 The Night Before

- ✅ **Sleep early** — your brain learns while you sleep!
- ✅ **Pack your supplies** — pencils, eraser, scratch paper, all ready to go.
- ✅ **Tell yourself:** "I've been practicing. I'm going to do great!"

👍 5 Simple Rules for Every Test

1. **Read the question twice.** The first time to understand it. The second time to catch details.
2. **Show your work.** Write the steps down, even on scratch paper. It helps you think!
3. **Skip the hard ones.** Put a small star next to tricky questions and come back later. Answer the easy ones first!
4. **Never leave a blank.** For multiple choice, your best guess is better than no answer at all.
5. **Check your work.** Finished early? Go back and re-read your answers.

✅ Smart Moves

- Take a deep breath before you begin
- Underline key words in the question
- Use drawings or number lines to help
- Cross out wrong answers first
- Double-check addition and subtraction

❌ Traps to Avoid

- Rushing and not reading carefully
- Picking the first answer that "looks right"
- Forgetting to carry or borrow numbers
- Skipping a question permanently
- Panicking when you see a tough problem

> 66 Remember, the very first practice test is the hardest — not because the questions are harder, but because everything is new! By Test 3, you'll feel like a pro. Trust me! 99

 Get Ready to Practice 

Pencils

Sharpened and ready!

Eraser

Everyone makes mistakes!

Scratch Paper

For working things out

A Calm Spot

Somewhere quiet to focus

A Grown-Up

To help set a timer

A Can-Do Attitude

You've totally got this!

✔ Allowed During Tests

- Pencils and erasers
- Blank scratch paper
- The **reference pages** in this book
- A ruler (for measurement questions)

✖ Not Allowed

- Calculators
- Phones, tablets, or computers
- Help from anyone else
- Your study guide (save it for after!)

For Parents & Teachers

- With only 3 tests, **space them at least a week apart.** This gives time to review mistakes before trying the next one.
- Let your child take Test 1 untimed to build familiarity.
- After each test, go through the Answer Key together. Focus on **understanding the "why,"** not just the score.
- If a topic keeps tripping them up, review it in our **Grade 3 Math Study Guide** before the next practice test.
- Celebrate every bit of progress — even getting one more question right is a win!

X^1 Math Reference Sheet X^1

You may use this page during your practice tests!

Symbol	Name	What It Means	
$+$	Plus (Add)	Put numbers together.	$3 + 5 = 8$
$-$	Minus (Subtract)	Take away from a number.	$9 - 4 = 5$
$\times$	Times (Multiply)	Add equal groups.	$4 \times 3 = 12$
$\div$	Divide	Split into equal groups.	$12 \div 3 = 4$
$=$	Equals	Both sides are the same.	$2 + 3 = 5$
$>$	Greater Than	The left number is bigger.	$7 > 3$
$<$	Less Than	The left number is smaller.	$2 < 9$
$\frac{1}{2}$	Fraction Bar	Part of a whole.	$\frac{1}{2}$ means 1 out of 2 equal parts

📖 Key Math Words

- **Sum** — the answer when you add
- **Difference** — the answer when you subtract
- **Product** — the answer when you multiply
- **Quotient** — the answer when you divide
- **Factor** — a number you multiply
- **Array** — objects in rows and columns
- **Fraction** — a part of a whole
- **Numerator** — the top number in a fraction
- **Denominator** — the bottom number
- **Equation** — a math sentence with $=$
- **Estimate** — a smart guess, close to the real answer
- **Perimeter** — the distance around a shape
- **Area** — the space inside a shape
- **Rounding** — making a number simpler by going to the nearest ten or hundred

🔍 Word Problem Clue Words

- **Add** (+): in all, total, altogether, combined, sum, both, more

- **Subtract** (−): how many more, how many left, fewer, difference, remain

- **Multiply** (×): each, every, groups of, times, rows of, per

- **Divide** (÷): share equally, split, each group, how many groups, per

Find more at
ViewMath.com/GA-Grade3

▦ Multiplication Table ▦

You may use this table during your practice tests!

×	1	2	3	4	5	6	7	8	9	10	11
1	1	2	3	4	5	6	7	8	9	10	11
2	2	4	6	8	10	12	14	16	18	20	22
3	3	6	9	12	15	18	21	24	27	30	33
4	4	8	12	16	20	24	28	32	36	40	44
5	5	10	15	20	25	30	35	40	45	50	55
6	6	12	18	24	30	36	42	48	54	60	66
7	7	14	21	28	35	42	49	56	63	70	77
8	8	16	24	32	40	48	56	64	72	80	88
9	9	18	27	36	45	54	63	72	81	90	99
10	10	20	30	40	50	60	70	80	90	100	110
11	11	22	33	44	55	66	77	88	99	110	121

♥ How to Use This Table

To find **4 × 7**:

1. Find **4** in the left column (blue).
2. Find **7** in the top row (blue).
3. Follow the row and column until they meet: the answer is **28**!

📈 My Confidence Tracker 📈

Record your scores below. You'll be amazed at your progress!

My name: ___________________________

✅ Test	📅 Date	⭐ Score	😊 How I Feel
1		/	
2		/	
3		/	

The easiest topic for me was:

____

The trickiest topic for me was:

____

One thing I got better at from Test 1 to Test 3:

____

Next time I want to try:

____

You just finished 3 practice tests — that's awesome! Compare your first score to your last. I bet you'll see real improvement. Ready for more? Check out our 5-test or 7-test books for even more practice!

 Table of Contents

Here's what we'll explore together!

⭐ Practice Test 1 1

⭐ Practice Test 2 9

⭐ Practice Test 3 17

⭐ Answer Key & Explanations 26

 Let's learn and have fun!

Practice Test 1

 *30 Questions*

✏️ Before You Start ✏️

- ✓ **Read each question carefully** before choosing your answer.
- ✓ **Show your work** on scratch paper when you need to.
- ✓ **Skip hard questions** and come back to them later.
- ✓ **Check your answers** when you're done.
- ✓ **Take your time** — there's no rush!

  You've Got This!

Do your best and show what you know!

1. Which number has a 4 in the hundreds place and a 7 in the thousands place?

   (A) 4,731

   (B) 7,431

   (C) 7,341

   (D) 3,470

2. What is 2,500 rounded to the nearest 1,000?

   (A) 2,000

   (B) 2,500

   (C) 3,000

   (D) 3,500

3. Which group contains only even numbers?

   (A) 12, 34, 57

   (B) 20, 46, 88

   (C) 31, 50, 72

   (D) 14, 63, 90

4. What is 7,777 rounded to the nearest 1,000?

   (A) 7,000

   (B) 7,700

   (C) 7,800

   (D) 8,000

5. What is $6,500 + 4,500$?

   (A) 10,000

   (B) 10,100

   (C) 11,000

   (D) 11,100

6. A farmer harvested 7,204 apples. He sold 3,568 apples at the market. How many apples does the farmer have left?

   Your Answer:

Find more at
ViewMath.com/GA-Grade3

7. *Estimate $248 + 175 + 362$ by rounding each number to the nearest hundred.*

Your Answer:

8. *In the number sentence $9 \times 6 = 54$, what is the product?*

(A) 9

(B) 6

(C) 54

(D) 15

9. *What is $8 \div 8$?*

Your Answer:

10. *What is $32 \div 4$?*

(A) 6

(B) 7

(C) 8

(D) 9

11. *A bakery makes 5 trays of muffins with 8 muffins each. Then they make 12 more muffins. How many muffins in all?*

(A) 25

(B) 40

(C) 52

(D) 60

12. *In an addition table, what pattern do you see going down the column for $+3$?*

(A) *Each number increases by 1*

(B) *Each number increases by 2*

(C) *Each number increases by 3*

(D) *Each number stays the same*

Find more at
ViewMath.com/GA-Grade3

13. You eat $\frac{2}{6}$ of a cake. How many equal parts does the whole cake have?

> Your Answer:

14. What whole number does $\frac{6}{3}$ equal?

(A) 1

(B) 2

(C) 3

(D) 6

15. You are at $\frac{3}{6}$ on a number line. You hop 2 more times by $\frac{1}{6}$. Where do you land?

> Your Answer:

16. $\frac{3}{4} = \frac{?}{8}$. What is the missing numerator?

(A) 3

(B) 4

(C) 6

(D) 7

17. Which of these fractions equals 0?

(A) $\frac{1}{1}$

(B) $\frac{0}{4}$

(C) $\frac{4}{4}$

(D) $\frac{4}{0}$

18. Is $\frac{2}{6}$ more or less than $\frac{1}{2}$?

> Your Answer:

Find more at
ViewMath.com/GA-Grade3

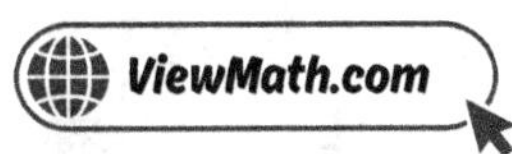

19. The minute hand points to the 12 and the hour hand points to the 5. What time is it?

(A) 12:05

(B) 12:25

(C) 5:12

(D) 5:00

20. 9,000 grams = how many kilograms?

(A) 9 kg

(B) 90 kg

(C) 900 kg

(D) 9,000 kg

21. A pitcher holds 8 liters of juice. The family drinks 3 liters. How much juice is left?

(A) 3 L

(B) 5 L

(C) 8 L

(D) 11 L

22. You have 3 quarters, 2 dimes, and 1 nickel. How much money do you have in cents?

Your Answer:

23. What is $2.50 + $1.25?

(A) $3.25

(B) $3.75

(C) $4.75

(D) $3.55

24. A picture graph has the key: Each ★ = 5 books. The row for Jake has 3 stars. How many books did Jake read?

(A) 3

(B) 5

(C) 8

(D) 15

Find more at
ViewMath.com/GA-Grade3

25. A bar graph has a scale that counts by 5s. A bar reaches up to the 4th line on the scale. What value does the bar show?

(A) 4

(B) 9

(C) 15

(D) 20

26. A line plot shows the lengths of 10 crayons. The data is shown below:

2 in.	$2\frac{1}{2}$ in.	3 in.	$3\frac{1}{2}$ in.	4 in.
2	3	4	1	0

How many crayons are shorter than 3 inches?

(A) 2

(B) 4

(C) 5

(D) 9

27. A shape has 4 sides. Two sides are 8 cm long and two sides are 3 cm long. It has 4 right angles. What shape is it?

Your Answer:

28. A square has sides that are 7 inches long. What is its area?

(A) 14 sq in

(B) 28 sq in

(C) 42 sq in

(D) 49 sq in

29. Lily says a rectangle that is 5 cm long and 3 cm wide has a perimeter of 15 cm. What mistake did she make?

(A) She subtracted instead of adding.

(B) She found the area instead of the perimeter.

(C) She forgot to add all 4 sides.

(D) She divided instead of multiplying.

Find more at
ViewMath.com/GA-Grade3

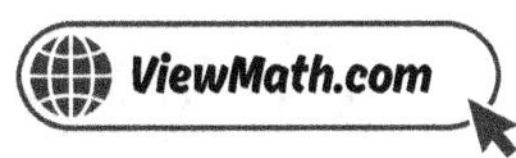

30. If one pizza is cut into 4 equal slices and another identical pizza is cut into 8 equal slices, which slices are bigger?

(A) The $\frac{1}{8}$ slices

(B) The $\frac{1}{4}$ slices

(C) They are the same size.

(D) You cannot tell.

 # ★ *End of Practice Test 1* ★ 

Great job finishing the test!

☑ *My Score*

I got ___________ out of 30 questions right.

*Check your answers in the **Answer Key** at the back of the book.*

💡 *Review any questions you missed. That's how we learn!*

📊 *Check Your Score Online!*

*Visit **ViewMath Academy** to enter your answers and see which topics you need to review. You can also explore lessons, take quizzes, track your scores, and save your progress!*

viewmath.com/score/3.1.GA.01

Or go to viewmath.com/score and enter code: 3.1.GA.01

2

Practice Test 2

☑ *30 Questions*

✏️ Before You Start ✏️

- ✓ **Read each question carefully** before choosing your answer.
- ✓ **Show your work** on scratch paper when you need to.
- ✓ **Skip hard questions** and come back to them later.
- ✓ **Check your answers** when you're done.
- ✓ **Take your time** — there's no rush!

★ You've Got This! ★

Do your best and show what you know!

1. A number has 3 thousands, 7 hundreds, 0 tens, and 5 ones. What is the number?

Your Answer:

2. What is 6,782 rounded to the nearest 1,000?

(A) 6,000

(B) 6,700

(C) 6,800

(D) 7,000

3. List all the even numbers between 21 and 30.

Your Answer:

4. A park had 6,238 visitors on Saturday and 4,715 on Sunday. Round each to the nearest 1,000 and estimate the total.

Your Answer:

5. What is $4,008 + 3,995$?

(A) 7,003

(B) 7,903

(C) 7,993

(D) 8,003

6. A school raised $7,500 for charity. They spent $4,825 on supplies. How much money is left?

Your Answer:

Find more at
ViewMath.com/GA-Grade3

7. Alex solved $345 + 287 = 632$. Is his answer reasonable?

   (A) Yes, it is reasonable

   (B) No, it should be closer to 500

   (C) No, it should be closer to 700

   (D) No, it should be closer to 800

8. Ms. Lee gives each of her 8 students 2 stickers. How many stickers does she give out in all?

   (A) 6

   (B) 10

   (C) 14

   (D) 16

9. What is $20 \div 4$?

   (A) 4

   (B) 5

   (C) 16

   (D) 24

10. A farmer has 63 eggs. He puts 7 eggs in each carton. How many cartons does he fill?

    (A) 7

    (B) 8

    (C) 9

    (D) 56

11. Noah has \$40. He buys 3 books that cost \$7 each. How much money does he have left?

    (A) \$12

    (B) \$19

    (C) \$21

    (D) \$33

12. Look at the pattern: $5, 10, 15, 20, 25$. What is the rule?

    (A) Add 3

    (B) Add 4

    (C) Add 5

    (D) Multiply by 2

Find more at
ViewMath.com/GA-Grade3

13. A pie is cut into 6 equal slices. 5 slices are left. What fraction of the pie is left?

(A) $\frac{1}{6}$

(B) $\frac{5}{6}$

(C) $\frac{6}{5}$

(D) $\frac{6}{6}$

14. Which fraction is closer to 1 on a number line divided into 6 parts?

(A) $\frac{2}{6}$

(B) $\frac{3}{6}$

(C) $\frac{4}{6}$

(D) $\frac{5}{6}$

15. Alex says $\frac{3}{4}$ is 3 copies of $\frac{1}{3}$. Is Alex correct?

(A) Yes

(B) No, $\frac{3}{4}$ is 3 copies of $\frac{1}{4}$

(C) No, $\frac{3}{4}$ is 4 copies of $\frac{1}{3}$

(D) No, $\frac{3}{4}$ is 3 copies of $\frac{3}{3}$

16. $\frac{1}{2} = \frac{?}{6}$. What is the missing number?

(A) 1

(B) 2

(C) 3

(D) 4

17. Write 4 as a fraction with denominator 3.

(A) $\frac{4}{3}$

(B) $\frac{3}{4}$

(C) $\frac{7}{3}$

(D) $\frac{12}{3}$

18. Is $\frac{5}{8}$ more or less than $\frac{1}{2}$? Explain.

Your Answer:

Find more at
ViewMath.com/GA-Grade3

19. Write the time for "quarter past 11" using numbers.

Your Answer:

20. A bag of rice has a mass of 5 kg. You use 2 kg for cooking. How much rice is left?

Your Answer:

21. Which container holds the LEAST liquid?

(A) A swimming pool

(B) A bathtub

(C) A bucket

(D) A teaspoon

22. Which group of coins equals 50 cents?

(A) 5 pennies

(B) 5 dimes

(C) 3 dimes

(D) 3 quarters

23. Lily has $6.00. She buys a sticker pack for $2.35 and a pen for $1.80. Does she have enough money left to buy a snack for $2.00?

(A) Yes, she has $2.85 left

(B) Yes, she has $2.00 left

(C) No, she only has $1.85 left

(D) No, she only has $1.65 left

24. A picture graph tracks books read in a month. Each symbol stands for 2 books. Amy has 5 symbols, Ben has 8 symbols, and Chloe has 3 symbols. How many books did they read altogether?

Your Answer:

Find more at
ViewMath.com/GA-Grade3

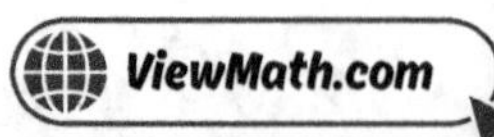

25. Using the same rainy days graph (Jan = 6, Feb = 4, Mar = 3, Apr = 7), how many fewer rainy days were in February than in April?

(A) 1
(B) 2
(C) 3
(D) 4

26. Leo measured 8 worms. His line plot shows 7 X marks total. What went wrong?

(A) He measured one worm twice
(B) He forgot to plot one measurement
(C) He used the wrong numbers on the number line
(D) Nothing — 7 is correct

27. A polygon has 6 sides. What is the name of this polygon?

Your Answer:

28. A square has sides of 9 inches. What is the area?

Your Answer:

29. A triangle has sides of 4 cm, 6 cm, and 8 cm. What is the perimeter?

(A) 14 cm
(B) 18 cm
(C) 20 cm
(D) 24 cm

30. You fold a piece of paper in half, then fold it in half again. How many equal parts do you have?

Your Answer:

Find more at
ViewMath.com/GA-Grade3

 End of Practice Test 2 

Great job finishing the test!

📋 My Score

I got __________ out of 30 questions right.

💡 *Review any questions you missed. That's how we learn!*

📊 Check Your Score Online!

Visit **ViewMath Academy** to enter your answers and see which topics you need to review. You can also explore lessons, take quizzes, track your scores, and save your progress!

viewmath.com/score/3.1.GA.02

Or go to viewmath.com/score and enter code: **3.1.GA.02**

Practice Test 3

30 Questions

✏️ Before You Start ✏️

- ✔ **Read each question carefully** before choosing your answer.
- ✔ **Show your work** on scratch paper when you need to.
- ✔ **Skip hard questions** and come back to them later.
- ✔ **Check your answers** when you're done.
- ✔ **Take your time** — there's no rush!

⭐ You've Got This! ⭐

Do your best and show what you know!

1. Which number is the same as 9 thousands, 0 hundreds, 2 tens, and 5 ones?

(A) 9,250

(B) 9,025

(C) 9,205

(D) 9,520

2. A store sold 423 books this month. About how many books is that, rounded to the nearest 100?

Your Answer:

3. If you add two odd numbers, the result is always:

(A) Odd

(B) Even

(C) Greater than 10

(D) An odd number less than 20

4. When rounding to the nearest 1,000, which digit do you look at?

(A) The ones digit

(B) The tens digit

(C) The hundreds digit

(D) The thousands digit

5. A plane flew 1,892 miles on Saturday and 2,347 miles on Sunday. How far did it fly in total?

Your Answer:

6. What is $4,000 - 1,637$?

(A) 2,263

(B) 2,363

(C) 2,463

(D) 3,363

Find more at
ViewMath.com/GA-Grade3

7. *Estimate $67 + 48$ by rounding each number to the nearest ten.*

   (A) 110   (B) 115

   (C) 120   (D) 130

8. *Which repeated addition matches 5×4?*

   (A) $5 + 4$ (B) $5 + 5 + 5 + 5$

   (C) $4 + 4 + 4 + 4 + 4$ (D) $4 + 4 + 4 + 4$

9. *What is $9 \div 1$?*

   (A) 0   (B) 1

   (C) 9   (D) 10

10. *What is $24 \div 3$?*

    (A) 6   (B) 7

    (C) 8   (D) 9

11. *Mom bakes 36 cookies and divides them equally onto 4 plates. Dad adds 3 cookies to each plate. How many cookies are on each plate now?*

    (A) 10   (B) 11

    (C) 12   (D) 13

12. *What is odd $\times$ even? Is the product odd or even?*

    Your Answer:

Find more at
ViewMath.com/GA-Grade3

13. A shape is divided into parts, but the parts are NOT equal sizes. Can you write a fraction for the shaded part?

(A) Yes, just count the shaded parts   (B) Yes, fractions work with any parts

(C) No, fractions only work when parts are equal   (D) No, you need to shade all the parts first

14. A number line is divided into 4 equal parts. Which fraction is at the halfway point between 0 and 1?

(A) $\frac{1}{4}$ (B) $\frac{2}{4}$

(C) $\frac{3}{4}$ (D) $\frac{4}{4}$

15. Count by fourths: $\frac{1}{4}$, $\frac{2}{4}$, __, $\frac{4}{4}$. What fraction is missing?

Your Answer:

16. $\frac{4}{8}$ simplified is:

(A) $\frac{1}{4}$ (B) $\frac{2}{4}$

(C) $\frac{1}{2}$ (D) $\frac{2}{8}$

17. Which of these fractions equals 1?

(A) $\frac{3}{4}$ (B) $\frac{6}{6}$

(C) $\frac{4}{6}$ (D) $\frac{1}{1}$ and $\frac{6}{6}$

18. Ruby says $\frac{1}{8} > \frac{1}{3}$ because $8 > 3$. Is Ruby correct?

(A) Yes, bigger numbers are always greater   (B) No, $\frac{1}{3} > \frac{1}{8}$ because thirds are bigger pieces than eighths

(C) They are equal   (D) Yes, eighths are bigger than thirds

Find more at
ViewMath.com/GA-Grade3

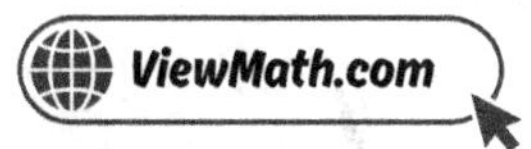

19. The short hand is past the 7 and the long hand is on the 10, then 2 more tick marks past the 10. What time is it?

   (A) 7:50   (B) 7:52

   (C) 10:37   (D) 7:10

20. A box of cereal has a mass of 500 g. You buy 2 boxes. What is the total mass?

   (A) 502 g   (B) 700 g

   (C) 1,000 g   (D) 5,000 g

21. A pot holds 6 liters when full. You pour in 2 liters first, then 3 more liters. How many more liters are needed to fill the pot?

Your Answer:

22. How much is a quarter worth?

   (A) 1 cent   (B) 5 cents

   (C) 10 cents   (D) 25 cents

23. What is $3.45 + $2.80?

   (A) $5.25   (B) $6.25

   (C) $5.65   (D) $6.15

24. A picture graph shows toys collected. Each symbol stands for 5 toys.

Bears: ★★★
Cars: ★★★★★
Dolls: ★★

How many toys were collected in all?

(A) 10   (B) 25

(C) 50   (D) 75

---

25. A bar graph has a scale that counts by 10s. The bar for Red reaches the 3rd line. How many items does Red show?

Your Answer:

---

26. Which type of graph is BEST for showing measurement data like the lengths of worms in inches?

(A) Picture graph   (B) Bar graph

(C) Line plot   (D) Tally chart

---

27. Which shape is both a rectangle AND a rhombus?

(A) Trapezoid   (B) Pentagon

(C) Square   (D) Triangle

---

28. A rectangle is 10 feet long and 3 feet wide. What is its area?

(A) 13 sq ft   (B) 26 sq ft

(C) 30 sq ft   (D) 33 sq ft

29. Which shape has the greatest perimeter?

(A) A square with sides of 5 cm

(B) A rectangle that is 8 cm × 3 cm

(C) A rectangle that is 6 cm × 4 cm

(D) A triangle with sides 7 cm, 7 cm, and 7 cm

30. A square is partitioned into 4 equal parts. All 4 parts are shaded. What fraction is shaded?

(A) $\frac{1}{4}$

(B) $\frac{3}{4}$

(C) $\frac{4}{4}$

(D) $\frac{4}{1}$

 # ★ *End of Practice Test 3* ★ 

Great job finishing the test!

☑ My Score

I got _____________ out of 30 questions right.

*Check your answers in the **Answer Key** at the back of the book.*

💡 *Review any questions you missed. That's how we learn!*

📊 Check Your Score Online!

Visit **ViewMath Academy** to enter your answers and see which topics you need to review. You can also explore lessons, take quizzes, track your scores, and save your progress!

viewmath.com/score/3.1.GA.03

Or go to viewmath.com/score and enter code: 3.1.GA.03

Answer Key & Explanations

⭐ Check Your Answers! ⭐

First try each test on your own, then look here to check.

Read the explanations to learn from any mistakes ⭐

☑ Practice Test 1 — Answer Key

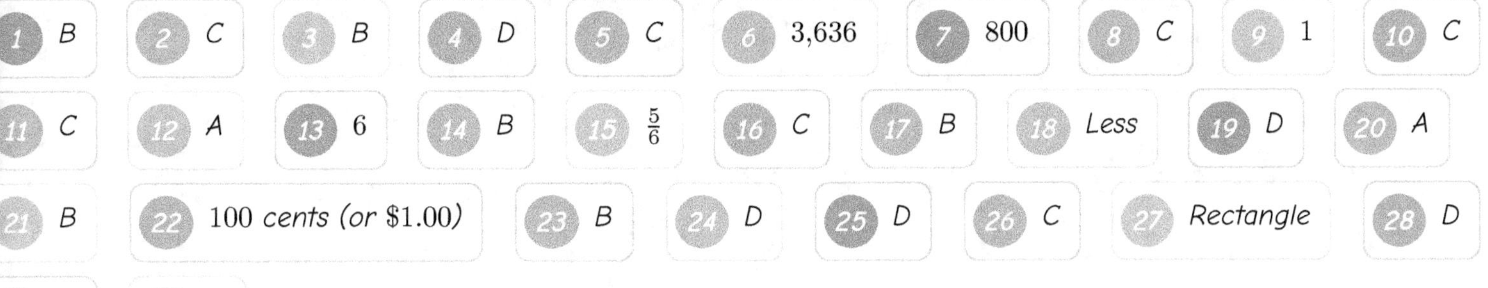

1 B	2 C	3 B	4 D	5 C	6 3,636	7 800	8 C	9 1	10 C
11 C	12 A	13 6	14 B	15 $\frac{5}{6}$	16 C	17 B	18 Less	19 D	20 A
21 B	22 100 cents (or $1.00)	23 B	24 D	25 D	26 C	27 Rectangle	28 D
29 B	30 B

💡 Time to Learn! 💡

Go through the explanations below, **especially for the questions you missed**.

Understanding why each answer is correct makes you a stronger math thinker!

👍 **Tip:** Circle any questions you got wrong, then read their explanation carefully.

📖 Practice Test 1 — Detailed Explanations

1 In 7,431: the 7 is in the thousands place and the 4 is in the hundreds place.

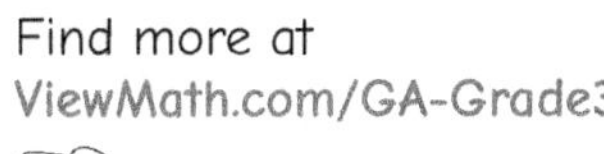

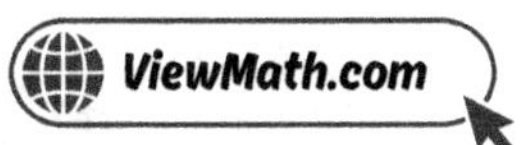

2   The hundreds digit is 5. Since $5 \geq 5$, round up. $2{,}500 \approx 3{,}000$.

3   20 (ends in 0), 46 (ends in 6), 88 (ends in 8) are all even. The other groups each contain at least one odd number.

4   The hundreds digit is 7. Since $7 \geq 5$, round up. $7{,}777 \approx 8{,}000$.

5 $6{,}500 + 4{,}500 = 11{,}000$. Two 4-digit numbers can sometimes add up to a 5-digit number!

6 $7{,}204 - 3{,}568 = 3{,}636$. Ones: $4 < 8$, borrow through zero: $14 - 8 = 6$. Tens: $9 - 6 = 3$. Hundreds: $1 - 5$, borrow: $11 - 5 = 6$. Thousands: $6 - 3 = 3$.

7 $248 \approx 200$, $175 \approx 200$, and $362 \approx 400$. So $200 + 200 + 400 = 800$.

8   The product is the answer when you multiply. In $9 \times 6 = 54$, the product is 54.

9   Any number divided by itself equals 1.

10 $4 \times 8 = 32$, so $32 \div 4 = 8$.

11   Step 1: $5 \times 8 = 40$ muffins. Step 2: $40 + 12 = 52$ muffins in all.

12   Going down a column in an addition table, as the row number increases by 1, each sum increases by 1.

13   The denominator tells the total equal parts. The cake is cut into 6 equal parts.

14 $\frac{3}{3} = 1$, so $\frac{6}{3} = 2$.

Find more at
ViewMath.com/GA-Grade3

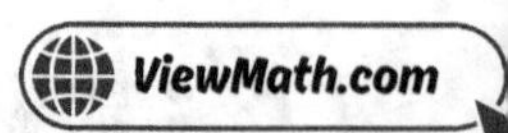

15 $\frac{3}{6} + \frac{1}{6} + \frac{1}{6} = \frac{5}{6}$.

16 The denominator went from 4 to 8 (multiplied by 2). So multiply the numerator by 2 too: $3 \times 2 = 6$.

17 When the numerator is 0, the fraction equals 0. $\frac{0}{4} = 0$.

18 $\frac{1}{2} = \frac{3}{6}$. Since $\frac{2}{6} < \frac{3}{6}$, it is less than $\frac{1}{2}$.

19 When the minute hand points to 12, it means 0 minutes (o'clock). The hour hand on 5 means the hour is 5. The time is 5:00.

20 Divide by 1,000: $9,000 \div 1,000 = 9$ kg.

21 $8 - 3 = 5$ L.

22 $3 \times 25 = 75$. Then $2 \times 10 = 20$. Then $1 \times 5 = 5$. Total: $75 + 20 + 5 = 100$ cents.

23 Line up the decimals and add: $\$2.50 + \$1.25 = \$3.75$.

24 Each star $= 5$ books. Jake has 3 stars, so $3 \times 5 = 15$ books.

25 The scale counts by 5s, so the 4th line is $4 \times 5 = 20$.

26 Crayons shorter than 3 inches: 2 in. (2) $+ 2\frac{1}{2}$ in. (3) $= 5$ crayons.

27 It has 4 right angles and 2 pairs of equal sides (but not all 4 sides are equal), so it is a rectangle.

28 A square is a rectangle with all sides equal. Area $= 7 \times 7 = 49$ sq in.

Find more at
ViewMath.com/GA-Grade3

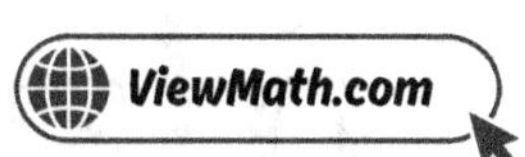

29 Lily multiplied $5 \times 3 = 15$, which gives the area, not the perimeter. The correct perimeter is $5 + 3 + 5 + 3 = 16$ cm.

30 The $\frac{1}{4}$ slices are bigger because the pizza is cut into fewer pieces. More equal parts means smaller pieces: $\frac{1}{4} > \frac{1}{8}$.

✅ Practice Test 2 — Answer Key

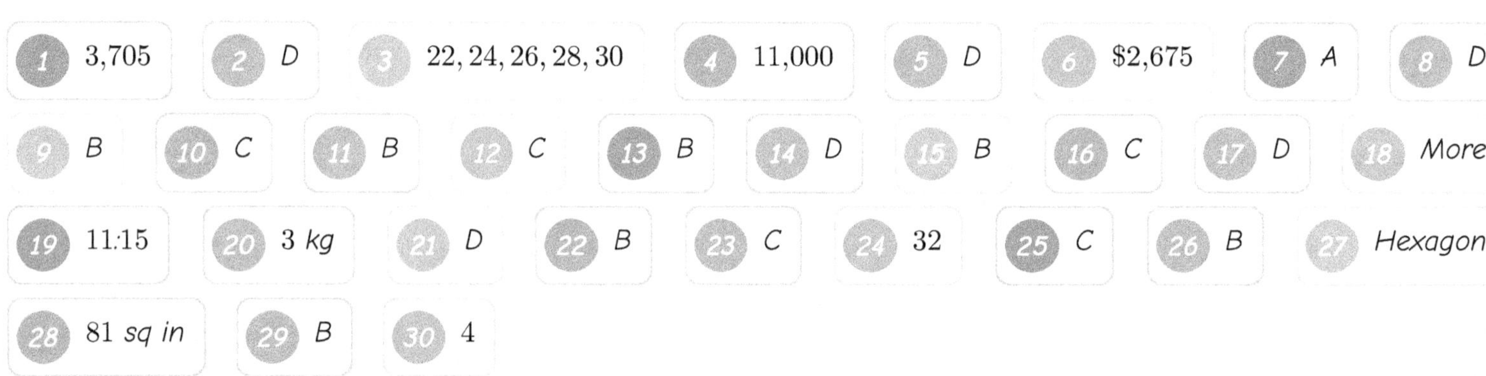

| 1 | 3,705 | 2 | D | 3 | $22, 24, 26, 28, 30$ | 4 | 11,000 | 5 | D | 6 | $2,675 | 7 | A | 8 | D |

9 B 10 C 11 B 12 C 13 B 14 D 15 B 16 C 17 D 18 More

19 11:15 20 3 kg 21 D 22 B 23 C 24 32 25 C 26 B 27 Hexagon

28 81 sq in 29 B 30 4

💡 Time to Learn! 💡

Go through the explanations below, **especially for the questions you missed.**

Understanding why each answer is correct makes you a stronger math thinker!

👍 **Tip:** Circle any questions you got wrong, then read their explanation carefully.

📖 Practice Test 2 — Detailed Explanations

1 3 thousands $= 3,000$, 7 hundreds $= 700$, 0 tens $= 0$, 5 ones $= 5$. The number is $3,000 + 700 + 5 = 3,705$.

2 The hundreds digit is 7. Since $7 \geq 5$, round up. $6,782 \approx 7,000$.

Find more at
ViewMath.com/GA-Grade3

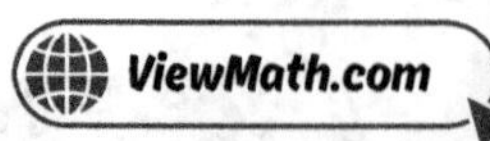

3 The even numbers between 21 and 30 are $22, 24, 26, 28, 30$. They all end in an even digit.

4 $6{,}238 \approx 6{,}000$ (hundreds digit $2 < 5$). $4{,}715 \approx 5{,}000$ (hundreds digit $7 \geq 5$). Estimated total: $6{,}000 + 5{,}000 = 11{,}000$.

5 Ones: $8 + 5 = 13$, carry 1. Tens: $0 + 9 + 1 = 10$, carry 1. Hundreds: $0 + 9 + 1 = 10$, carry 1. Thousands: $4 + 3 + 1 = 8$. The sum is $8{,}003$. Watch out for chains of carries through zeros!

6 $7{,}500 - 4{,}825 = 2{,}675$. Borrow as needed: ones $10 - 5 = 5$, tens $9 - 2 = 7$, hundreds $4 - 8$ requires borrowing: $14 - 8 = 6$, thousands $6 - 4 = 2$.

7 $345 \approx 300$ and $287 \approx 300$. The estimate is $300 + 300 = 600$. Alex's answer of 632 is close to 600, so it is reasonable.

8 8 students each getting 2 stickers is $8 \times 2 = 16$ stickers total.

9 20 split into 4 equal groups gives 5 in each group. $20 \div 4 = 5$.

10 $63 \div 7 = 9$ cartons.

11 Step 1: $3 \times \$7 = \$21$ spent. Step 2: $\$40 - \$21 = \$19$ left.

12 Each number is 5 more than the one before it. The rule is "add 5" (these are multiples of 5).

13 5 slices out of 6 equal slices $= \frac{5}{6}$.

14 $\frac{5}{6}$ is only $\frac{1}{6}$ away from 1, closer than all the other choices.

15 $\frac{3}{4}$ means 3 copies of $\frac{1}{4}$, not $\frac{1}{3}$. The denominator tells the size of each piece.

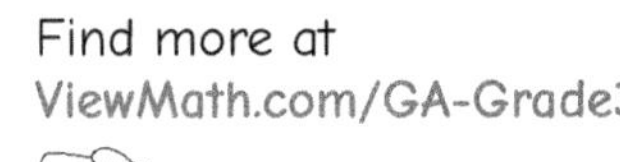
Find more at
ViewMath.com/GA-Grade3

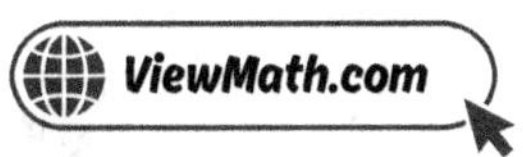

16 The denominator went from 2 to 6 (multiplied by 3). Multiply the numerator too: $1 \times 3 = 3$. So $\frac{1}{2} = \frac{3}{6}$.

17 $4 \times 3 = 12$ thirds. So $4 = \frac{12}{3}$.

18 $\frac{1}{2} = \frac{4}{8}$. Since $\frac{5}{8} > \frac{4}{8}$, it is more than $\frac{1}{2}$.

19 "Quarter past" means 15 minutes after the hour. Quarter past 11 is 11:15.

20 $5 - 2 = 3$ kg.

21 A teaspoon holds much less than 1 liter, which is far less than a bucket, bathtub, or pool.

22 5 dimes $= 5 \times 10 = 50$ cents.

23 Total spent: $\$2.35 + \$1.80 = \$4.15$. Money left: $\$6.00 - \$4.15 = \$1.85$. Since $\$1.85 < \$2.00$, she does not have enough.

24 Amy: $5 \times 2 = 10$. Ben: $8 \times 2 = 16$. Chloe: $3 \times 2 = 6$. Total: $10 + 16 + 6 = 32$ books.

25 $7 - 4 = 3$ fewer rainy days in February than April.

26 He measured 8 worms but only has 7 X marks, so he forgot to plot one measurement.

27 A polygon with 6 sides is called a hexagon. "Hex" means 6.

28 Area $= 9 \times 9 = 81$ sq in.

29 Add all side lengths: $4 + 6 + 8 = 18$ cm.

Find more at
ViewMath.com/GA-Grade3

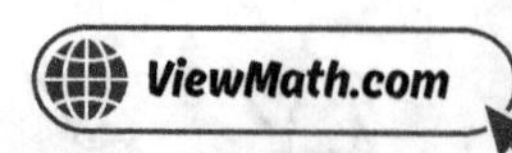

30 Folding in half gives 2 parts. Folding in half again doubles it to 4 equal parts. Each part is $\frac{1}{4}$ of the whole.

✅ Practice Test 3 — Answer Key

1 B	**2** 400	**3** B	**4** C	**5** 4,239	**6** B
7 C	**8** C	**9** C	**10** C		

11 C	**12** Even	**13** C	**14** B	**15** $\frac{3}{4}$	**16** C
17 D	**18** B	**19** B	**20** C		

21 1 L	**22** D	**23** B	**24** C	**25** 30	**26** C
27 C	**28** C	**29** B	**30** C		

> ### 💡 Time to Learn! 💡
>
> Go through the explanations below, **especially for the questions you missed.**
>
> Understanding why each answer is correct makes you a stronger math thinker!
>
> 👍 **Tip:** Circle any questions you got wrong, then read their explanation carefully.

📖 Practice Test 3 — Detailed Explanations

1 9 thousands $= 9,000$, 0 hundreds $= 0$, 2 tens $= 20$, 5 ones $= 5$. So $9,000 + 20 + 5 = 9,025$.

2 The tens digit is 2. Since $2 < 5$, round down. $423 \approx 400$.

3 Odd + Odd = Even. For example, $3 + 5 = 8$ (even) and $7 + 9 = 16$ (even).

4 To round to the nearest 1,000, look at the hundreds digit. If it is 5 or more, round up. If it is 4 or less, round down.

Find more at
ViewMath.com/GA-Grade3

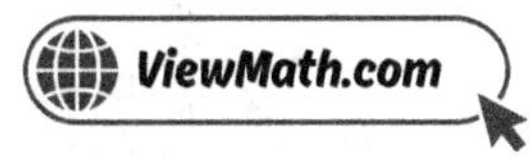

5 $1{,}892 + 2{,}347 = 4{,}239.$ *Ones:* $2 + 7 = 9.$ *Tens:* $9 + 4 = 13$, *carry 1. Hundreds:* $8 + 3 + 1 = 12$, *carry 1. Thousands:* $1 + 2 + 1 = 4.$

6 *Borrow through all the zeros: thousands* $4 \to 3$, *hundreds* $0 \to 9$, *tens* $0 \to 9$, *ones* $0 \to 10$. *Then* $10 - 7 = 3$, $9 - 3 = 6$, $9 - 6 = 3$, $3 - 1 = 2$. *The answer is* $2{,}363$.

7 $67 \approx 70$ *and* $48 \approx 50$. *So* $70 + 50 = 120$.

8 5×4 *means 5 groups of 4, which is* $4 + 4 + 4 + 4 + 4 = 20$.

9 *Any number divided by 1 equals itself.* $9 \div 1 = 9$.

10 $3 \times 8 = 24$, *so* $24 \div 3 = 8$.

11 *Step 1:* $36 \div 4 = 9$ *cookies per plate. Step 2:* $9 + 3 = 12$ *cookies per plate.*

12 *Odd* $\times$ *even always equals an even number. Example:* $3 \times 4 = 12$.

13 *Fractions only work when the parts are* **equal**. *If the pieces are different sizes, you cannot write a fraction.*

14 $\frac{2}{4}$ *is at the second of 4 marks, which is the middle of the line from 0 to 1.*

15 *Counting by* $\frac{1}{4}$: $\frac{1}{4}, \frac{2}{4}, \frac{3}{4}, \frac{4}{4}$.

16 *Divide top and bottom by 4:* $\frac{4 \div 4}{8 \div 4} = \frac{1}{2}$.

17 *Both* $\frac{1}{1}$ *and* $\frac{6}{6}$ *equal 1 because numerator = denominator.*

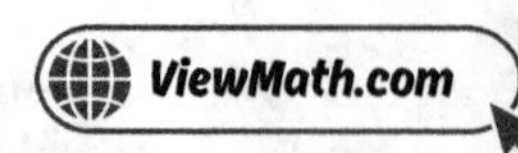

18 With the same numerator, the bigger denominator gives **smaller** pieces. $\frac{1}{3} > \frac{1}{8}$.

19 The hour is 7. The long hand on 10 means 50 minutes. Two more tick marks gives $50 + 2 = 52$ minutes. The time is 7:52.

20 $500 + 500 = 1,000$ g (which is the same as 1 kg).

21 You poured in $2 + 3 = 5$ L. The pot holds 6 L, so $6 - 5 = 1$ more liter is needed.

22 A quarter is worth 25 cents.

23 Cents: $45 + 80 = 125$ cents $= 1$ dollar and 25 cents. Dollars: $3 + 2 + 1 = 6$. Answer: $6.25.

24 Bears: $3 \times 5 = 15$. Cars: $5 \times 5 = 25$. Dolls: $2 \times 5 = 10$. Total: $15 + 25 + 10 = 50$ toys.

25 The scale counts by 10s, so the 3rd line is $3 \times 10 = 30$ items.

26 Line plots are best for measurement data. They show each individual measurement as an X on a number line, making it easy to see how values are spread out.

27 A square has 4 right angles (like a rectangle) AND 4 equal sides (like a rhombus). So a square is both a rectangle and a rhombus.

28 Area $= 10 \times 3 = 30$ sq ft.

29 Square: $4 \times 5 = 20$ cm. Rectangle 8×3: $(2 \times 8) + (2 \times 3) = 22$ cm. Rectangle 6×4: $(2 \times 6) + (2 \times 4) = 20$ cm. Triangle: $7 + 7 + 7 = 21$ cm. The 8×3 rectangle has the greatest perimeter of 22 cm.

30 All 4 out of 4 parts are shaded, so $\frac{4}{4}$ is shaded. $\frac{4}{4} = 1$ whole.

Find more at
ViewMath.com/GA-Grade3

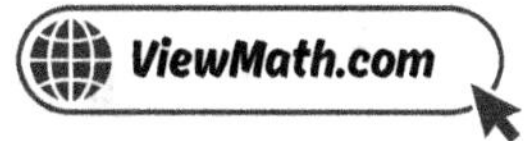

Great job checking your work!

Keep practicing and you'll be a math star!

Find more at
ViewMath.com/GA-Grade3